CATS

Ancient and Modern

CATS

Ancient and Modern

Juliet Clutton-Brock

Harvard University Press
Cambridge, Massachusetts
1993

Acknowledgements

I am indebted to Professor P.P.G. Bateson, F.R.S., Sub-Department of Animal Behaviour, University of Cambridge, for his helpful comments on the socialisation of kittens. I am also most grateful to Daphne Hills, of the Mammal Section at the Natural History Museum, for information on the feral and hybrid cats of Scotland.

My thanks are due to Rachel Rogers, of British Museum Publications, who was responsible for the picture research. The manuscript was typed by Jackie Burnham.

The author and publishers are also grateful to the following for permission to reproduce illustrations:

Animal Photography 44, 58, 59, 63, 65, 73, 79 *bottom*, 81, 82, 83; Animals Unlimited 76 *top*, 77 *bottom*; Ardea 68 *bottom*; BBC Hulton Picture Library 51, 53; Bodleian Library 48; British Library 41, 43, 46 *top*, 47, 50; British Museum half-title page, frontispiece, title page, 6, 7, 8, 9 *bottom*, 17 *top*, 18, 20, 21, 22, 25, 26, 27, 28, 29, 32 *both*, 33, 34–5, 36, 39 *bottom*, 40 *right*, 54, 60, 63, 69 *top*, 71, 78, 80, 84 *bottom*, 86, 88, 92; British Museum (Natural History) 9 *top*, 14, 37 *both*, 57, 70 *top*, 79 *top*, 89; Bruce Coleman Ltd 23; Fitzwilliam Museum, Cambridge 16; Harry Green 24; Marc Henrie 62, 67 *bottom*, 69 *bottom*, 70 *bottom*, 76 *bottom*, 77 *top*, 84 *top*; Deborah and Bruce Jackson 56; National Archaeological Museum, Athens 39 *top*; Oxford Scientific Films Ltd 11, 12, 13 *top*; Reading Museum and Art Gallery 30; Rex Features Ltd 93 *top*; Oriel Robinson 72; Royal Commission on the Historical Monuments of England 49; Royal Naval Museum 91; RSPCA 88 *bottom*; Staatliche Museen, Berlin 52; Topham Picture Library 87; Trinity College Library, Dublin 42; Victoria and Albert Museum 95; Zoological Society of London 13 *bottom*

Printed in Hong Kong
10 9 8 7 6 5 4 3 2 1

Library of Congress Cataloging-in-Publication Data
Clutton-Brock, Juliet.
  Cats, ancient and modern/Juliet Clutton-Brock.
    p.   cm.
  Includes bibliographical references and index.
  ISBN 0-674-10407-2
  1. Cats—History.   2. Cats.   I. Title.
SF442.6.C58 1993
636.8′009—dc20   93-19128
                    CIP

Half title Curious long-haired cat from *Our Cats* by Harrison Weir, 1889.

Frontispiece Cat sleeping under peonies, mid-19th century. Japanese hanging scroll, ink and colours on silk. (British Museum)

Title Vignette of a little girl holding an angry cat by Francesco Bartalozzi (1728–1815). Pen and ink. (British Museum)

Contents Striped tabby. Drawing by Théophile Steinlen (1859–1923).

Contents

ACKNOWLEDGEMENTS   4

THE CAT'S PLACE IN NATURE   6

Wild relatives   6

Reproduction   15

Behaviour   17

CATS IN ANCIENT TIMES   26

Early domestication   26

Ancient Egypt   36

The Classical world   39

The Middle Ages   41

THE CAT IN LEGEND AND WITCHCRAFT   51

CATS IN MODERN TIMES   59

The foundation breeds   59

    Manx   62

    Siamese   64

    Abyssinian   67

    Angora or Persian   68

    Russian Blue   70

The Cat Fancy   71

Distribution of the common cat in town and country   79

Cats in the Orient   83

Feral cats   87

Ships' cats   90

REFERENCES AND FURTHER READING   94

INDEX   96

The Cat's Place in Nature

Above Belled cat by Hiroaki Takahashi (1871–1944). Japanese woodblock print, *c*. 1935. (British Museum)

Opposite Ivory *netsuke* tiger. Japanese, 19th century. (Hull Grundy Collection, British Museum)

WILD RELATIVES

It is often said that the domestic cat is an exploiter of humans rather than the other way round. This is because the cat's solitary nature, demanding personality and secret night life sets it apart from all other domestic animals. Nevertheless, the cat may be even more popular as a household companion than the dog, and a recent estimate claimed that there are as many as 400 million cats in the world today. All these belong to one species which was named *Felis catus* by Linnaeus in 1758.

There are twenty-six species of small wild cats belonging to the same group or genus as the domestic cat, being all named *Felis*, and these, together with three species of *Lynx*, six species of large cats (*Panthera*), and the cheetah (*Acinonyx*), make up the family of the Felidae. With the exception of the cheetah, the felids are a contiguous group of carnivores which are all recognisable as belonging to one mammalian family. They are all long-legged, short-faced, and have lithe bodies with sleek coats. They also have retractile claws, highly developed senses, and they purr. All the felids, except the lion, are solitary hunters that kill rather small prey and they are mostly nocturnal. The lion is the most gregarious of the felids and it usually lives and hunts in a family group known as the pride.

The domestic cat is the only felid that lives and breeds within human societies, although the cheetah can be tamed and has been used in the Middle East and India in the hunting of gazelle for 1,000 years or more.

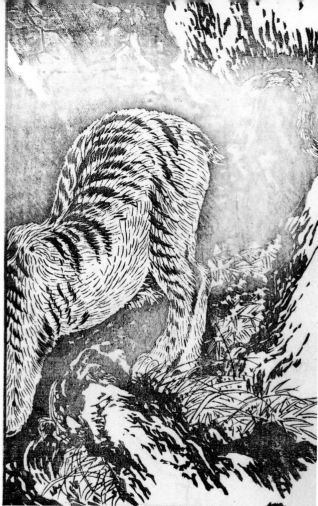

A wild cat from *Suiseki Gafu* ('Suiseki's Picture Album') by Sato Suiseki (worked *c*. 1806–40). Japanese woodblock album. (British Museum)

The species of cats within the same genus as the domestic cat (*Felis*) are found as indigenous wild animals throughout the world except in Australasia, some Oceanic Islands and the Antarctic. The domestic cat is found in all places where humans have travelled, and it is probably even more widespread than the dog because of its capacity to survive on its own as a feral animal – that is, one that has returned to live and breed in the wild.

Of the twenty-six species of small wild cats there is only one that has a serious claim to being the ancestor of the domestic cat. This is the very widely distributed wild cat of Europe, the Middle East and Africa. In the north this species is typified by the Scottish wild cat; in the south it is known as the African wild cat. These two races form the extremes of the range and are so different in appearance that they have often been described as separate species. However, they have a continuous distribution and will readily interbreed so that it is more appropriate, biologically, to consider them as a single species. Throughout its range the wild cat has a striped tabby coat but it is more heavily marked in the north and more flecked in the south, while the Scottish wild cat has a relatively heavy body and head, rather stocky limbs and short ears. The African wild cat is more fine-limbed, has larger ears and a smaller head, these characteristics being commonly associated with mammals from cold and hot climates respectively.

Since the time of Linnaeus the Latin name, *Felis catus*, has been commonly used for the European wild cat as well as for the domestic breeds. It is now generally agreed, however, that the wild species should not be described by the same scientific name as the domestic form unless they are indistinguishable from it. The name *Felis catus* is particularly inappropriate for the wild cat because this species has a coat that is striped, not blotched, as described by Linnaeus who had probably never seen a wild cat and may have been describing his own blotched tabby. This was pointed out by the well-known taxonomist of mammals R. I. Pocock in 1907, and since his time it has been usual to name the wild cat *Felis silvestris*, as first used by Schreber in 1777, and the domestic cat *Felis catus* Linnaeus, 1758.

Scottish wild cat, *Felis silvestris*.

Left Coloured print of a striped tabby cat by George Baxter (1804–67). (British Museum)

Overleaf Blotched tabby playing with a dead bird.

Page 11 African wild cat, *Felis silvestris libyca*. From J. Anderson, *Zoology of Egypt: Mammalia*, London, 1902.

[9]

Of all domestic cats the striped tabby is the one that most closely resembles the wild *Felis silvestris* and it is therefore reasonable to assume that this is the foundation type. Although not much used at the present day, Pocock coined the name *torquata* for the striped tabby and defined its coat as, 'Sides of the body, from the shoulder to the root of the tail, marked with narrow wavy vertical stripes which show a tendency, especially on the thighs, to break up into spots; no broad latero-dorsal stripe'.

For the second, blotched type of tabby Pocock used the name *catus* and described the coat as, 'Sides of the body marked with three usually obliquely longitudinal stripes forming the so-called "spiral", "horseshoe", or "circular" pattern of fanciers; a broad latero-dorsal stripe on each side of the narrow median spinal stripe'.

The northern race of wild cat is nowadays named *Felis silvestris silvestris* and the southern race *Felis silvestris libyca*. It is this race and the western Asiatic *Felis silvestris ornata* that are generally assumed to be the ancestors of the domestic cat. Other potential progenitors that have been considered are Pallas's cat and the sand cat, both of which have been suggested as ancestors of the domestic Persian cat, and the Asian golden cat and leopard cat as ancestors of the Siamese.

Pallas's cat, *Felis manul*, is found from Iran to west China in mountain and steppe country, and the sand cat, *Felis margarita*, is a desert animal found from the Sahara to Baluchistan. The golden cat, *Felis temminckii*, is a forest species occurring from south China to Sumatra, while the leopard cat, *Felis bengalensis*, has a more widespread distribution from eastern Siberia through Baluchistan to South-east Asia.

The general appearance of these wild cats has led to suggestions that they could have been the ancestors of the Persian and Siamese cats, but detailed studies of the skulls and skeletons of the wild species in comparison with the domestic breeds by a number of biologists, notably Kratochvíl and Hemmer, have shown that there is no anatomical foundation for the suppositions. On the other hand, all breeds of domestic cats are found to bear a very close anatomical resemblance to the North African and Asiatic forms of the wild cat, *Felis silvestris*, belonging to the subspecies *libyca* and *ornata*. There is support for this anatomical evidence in the chromosome number of the domestic cat and these subspecies of wild cat (as well as *Felis bengalensis* and *Felis manul*), in all of which there are nineteen pairs of chromosomes.

Apart from the close similarity in the colour and pattern of the coat in the wild cat and in the domestic striped tabby (the most 'primitive' domestic cat), it has been proved by Hemmer that the size of the brain, measured as the cranial capacity of the skull, is very closely similar in the domestic cat and in the *libyca* and *ornata* races of the wild cat, *Felis silvestris*. Hemmer has been carrying out research for many years at the University of Mainz in West Germany on the relative brain size of different groups of domestic mammals and he has found that it is always smaller in domesticates than in the wild species. He has also found that within the wild species the size of the brain is smaller in the race that is the direct ancestor of the domestic species. This is so with the wild cat where the brain of the southern races is relatively smaller than in the north European wild cat.

Above Sand cat, *Felis margarita.*

Left Pallas's cat, *Felis manul.*

Far left Golden cat, *Felis temminckii.*

REPRODUCTION

Cats are prolific breeders: the female can have kittens from the time she is little more than six months old. The male matures rather later, but both can breed before they are a year old. The size of the litter usually varies from three to ten in number and the kittens are weaned by the time they are eight weeks old. The mother can come into heat (oestrus) again four weeks after weaning. The mating season depends on day length: in northern Europe it is usual for the domestic female to come into heat in January and again in June.

A few days before she is due to give birth the domestic mother cat will become restless and will try to find a warm, dark nest in some quiet place. The kittens are born blind but are relatively well-developed, being more responsive than newborn puppies. Each kitten will soon claim its own teat and will suck at no other, even if another kitten dies leaving a teat full.

Above Tabby mother with her kittens, showing the variable coat colours of the European domestic cat. From D.G. Elliot, *A Monograph of the Felidae, or Family of the Cats*, London, 1853, pl. 41.

Opposite Indian desert cat, *Felis silvestris ornata*. From T. Hardwicke (1756–1835), *Original drawings*, vol. I, pl. 42. (Natural History Museum)

Sketches of a kitten by Alexandre-François Desportes (1661–1743). (Fitzwilliam Museum, Cambridge)

The social life of the cat begins immediately after birth, and as with all domestic animals the way it is habituated to humans from the very beginning will greatly affect its ability to be an animal companion in adult life. Many pedigree cats appear to be 'neurotic' in their behaviour, and this is claimed to be because they are 'highly-bred', but it is much more likely that their nervousness stems from a lack of human contact in their first weeks. Many pedigree cats are reared in a cattery and are not handled sufficiently by a sympathetic human when they are in the important early period of socialisation. The kitten needs to be gently held, stroked and talked to from the time that its eyes open when it is about ten days old. It will begin to play when it is about three weeks old and will be weaned by the time it is six weeks; from this time it should become increasingly aware of its place in the home of its human family.

Studies of the Scottish wild cat indicate that it has a similar pattern of reproduction. The females come into heat at the begining of March, and the kittens are born in the second or third week of May after a gestation of sixty-eight days. The mother may then come into heat again at the end of May when the kittens are weaned. If a second litter is born, it will be during August, although a second mating may be irregular in occurrence and breeding may not take place at all in wild cats kept in captivity or in domestic cats that are closely confined. Very occasionally wild cats have been known to breed three times in a year.

BEHAVIOUR

Although both are carnivores, cats are quite different from dogs. The dog is the most social of all domestic animals, and its inherited behaviour patterns closely resemble those of humans which is why the dog can be so closely integrated into the home

Above The Girl and Kitten, coloured print (published 1787), after Sir Joshua Reynolds (1723–92). (British Museum)

Left Kittens' first encounter with a mouse, from *Our Cats* by Harrison Weir, 1889.

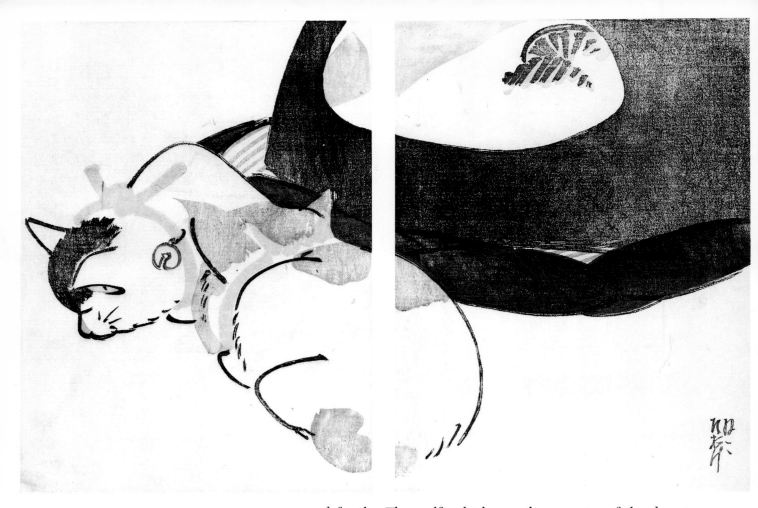

and family. The wolf, which was the ancestor of the dog, is a
social hunter using communal efforts to hunt prey animals that
are larger than itself. The wolf family forms a closely united pack
in which there are dominant and submissive individuals who
maintain a strict hierarchy. Male and female wolves may mate
for life, and both parents help with feeding and care of the cubs.
The wolf may have a huge home range, the size depending on
the amount of prey available, and it will travel over great
distances in search of food. In all these ways, and many more,
the behaviour of the wolf, and hence the dog, is unlike that of
the cat.

The cat is a solitary hunter, 'walking by his wild lone through
the wild wet woods and waving his wild tail', as described by

Kipling in his allegory on the history of domestication in the *Just So Stories*. Kipling was wrong, however, in claiming for the cat 'that all places are alike to him', for cats are strongly territorial. He would have been more accurate, perhaps, if he had read the description of the cat by Topsell who wrote in 1607:

The nature of this beast is to love the place of her breeding; neither will she tarry in any strange place, although carried far. She is never willing to forsake the house for love of any man, and this is contrary to the nature of a dog, who will travel abroad with his master. Although their masters forsake their houses, yet will not cats bear them company, and being carried forth in close baskets or sacks, they will return again or lose themselves.

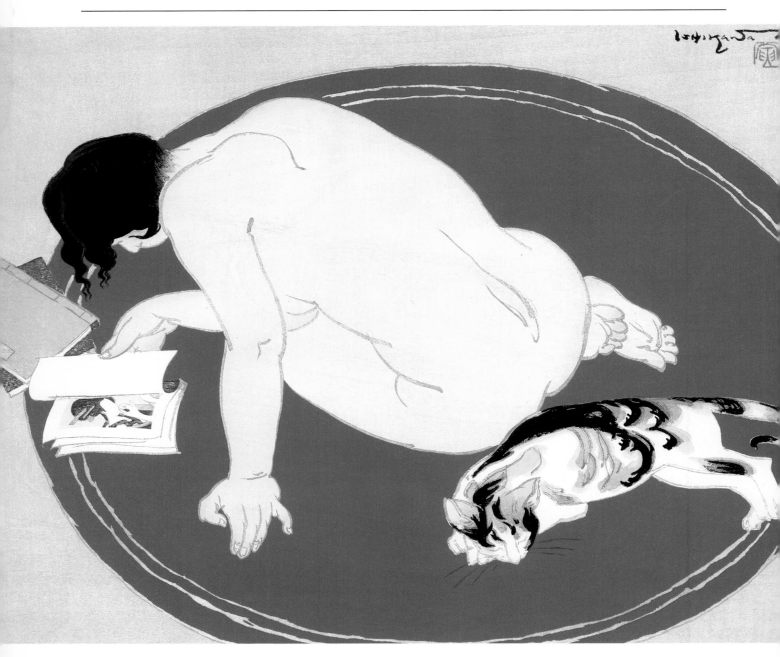

As with all behaviour the inherited patterns are not simple or clear-cut, and although the cat is a solitary hunter (only one individual is needed to kill a mouse or a small bird) it is social in some of its activities. The cat will share its core area with its human family and it will sleep either with people or with other cats, primarily perhaps for warmth.

Like all felids the domestic cat has a wide repertoire of sounds, of which purring is perhaps the only one that is attractive to the human ear. We think of the purr as being a sign of contentment in a cat that is lying cosily on a lap, in a kitten that is sucking from its mother, or in a mother that is suckling her kittens; but cats will also purr when they are in pain. They will usually greet members of their group, either human or animal, with a miaow or a chirrup, and they may make a clicking noise when they are chasing something. Noises concerned with sexual behaviour in cats include the caterwaul of the male, the growling-moaning call of the female, and the scream of the mated female.

The cleanliness of the cat family makes these animals particularly suited to living with humans. Cats are very easy to 'house-train' because their natural instinct is to urinate and defecate away from their immediate home territory and to bury their excreta. The tongue of the cat is peculiarly 'rough', and most cats will keep themselves scrupulously clean by licking themselves all over.

The body of the dog is built for speed and the ability to range over great distances in pursuit of prey that is bigger than itself. The body of the cat is built for agility, climbing, pouncing, jumping and short bursts of speed after prey that is smaller than itself. This helps to explain the very different behaviour of the domestic dog and cat: a dog can be trained to hunt in a pack, or herd sheep, or race against other dogs; a cat can be very little trained but it enjoys the proverbial 'nine lives' which result from its extremely quick and agile physical reactions to unexpected events.

Studies of feral cats living in the wild, as discussed in *The Wildlife of the Domestic Cat* by Roger Tabor (1983), have shown that the household cat differs little in its behaviour from those living wild. The human home is its core area where it feeds, sleeps and gives birth. The home range is contained by the

Above Cat licking its paw, attributed to Katsukawa Shunshō (1726–92). Japanese hanging scroll, ink and colours on paper. (British Museum)

Opposite Leisure hours by Toraji Ishikawa (1875–1964). Japanese woodblock print showing nude with soundly sleeping cat, 1934. (British Museum)

Right Cat looking at fields at Asakusa by Andō Hiroshige (1797–1858). Japanese woodblock print from *100 Views of Edo*. (British Museum)

Far right Cat jumping off a shelf in slow motion.

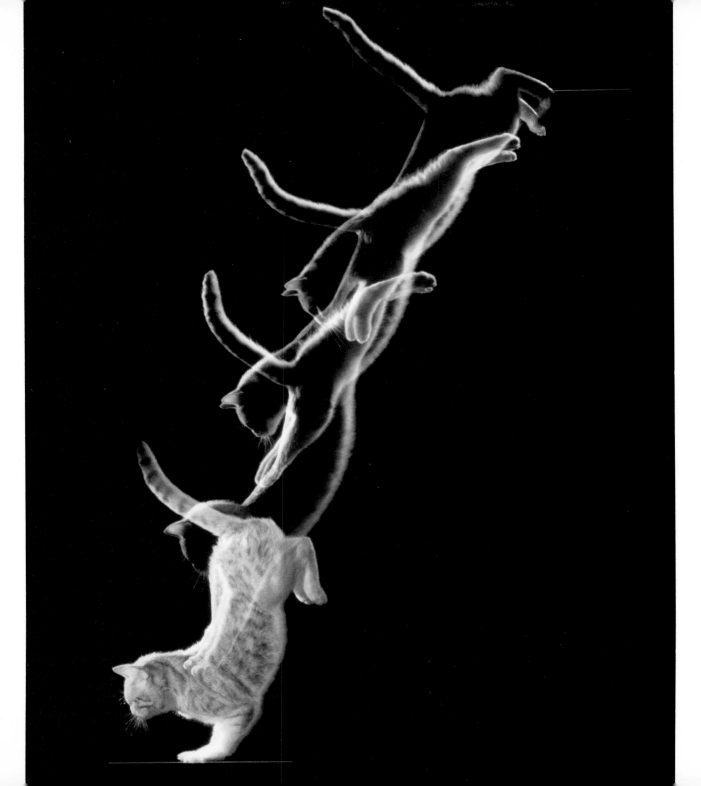

garden boundaries and perhaps by that part of the road where the family's car is parked, and within this area there is a strongly defended territory which is larger for the male than the female. Cats are very watchful and they are quite capable of knowing which hedge, lawn, driveway or motor car belongs to their own human family.

As Tabor points out, the cat is the only household animal (unless the mouse is included) that is entirely free-ranging and therefore its breeding is uncontrolled by humans. This has meant that over the millennia since the cat was first domesticated it has undergone very little artificial selection for characteristics favoured for aesthetic or subjective reasons, and breeds of cats were not developed until the last century. There must, however, have been a certain amount of selection for coat colour as people have very definite views on what colour cats they like best. Recent work by Hemmer and others has shown that there is a link between temperament and coat colour. It appears probable that mutant coat colours, such as black, which differ from the wild type (the striped tabby), have a genetic link with a temperament that is less timid and will tolerate higher population densities. This has an obvious advantage for the urban cat of today (see p.80).

Left Sketch of a cat by Gwen John (1876–1939). Pencil, watercolour and wash. (British Museum)

Far left Striped tabby kitten perched on the boot of his family's motor car.

[25]

Cats in Ancient Times

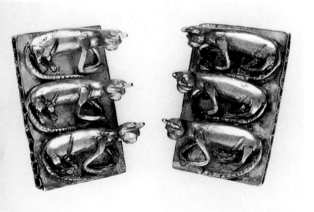

Above Two spacer-bars, each with three cats. Gold, from a bracelet, possibly from the tomb of Queen Sobkemsaf at Edfu, Egypt, *c*. 1650 BC. (British Museum)

Opposite Cat hunting fowl in the marshes. Detail from a wall-painting from the tomb of Nebamun, Thebes (*c*. 1400 BC). (British Museum)

EARLY DOMESTICATION

It is not known when cats were first domesticated. The bones and parts of the skulls of cats are quite frequently found at archaeological sites from the early prehistoric period. Presumably these are from wild animals killed for their furs as they are usually together with the bones of other wild carnivores such as wolf, jackal, fox, lynx or bear. Probably, however, there were tamed kittens that grew into adult cats and stayed near to human habitations long before their remains appear in the archaeological record.

The remains of cats have been found from a number of early sites of note in different parts of the world. These include Jericho (now in Israel) where a cat's tooth was identified from the pre-pottery Neolithic level dating to around 9,000 years ago, and another from Harappa in the Indus valley from around 4,000 years ago. More importantly for the history of the cat's relationship with humans, remains of a cat have recently been identified from the pre-pottery Neolithic levels of a site in Cyprus dated to around 7,000 years ago. The significance of this find lies in the fact that this cat (or its forebears) must have been taken to Cyprus, because there is no fossil record for the presence of the cat in Cyprus before the first human immigration to this Mediterranean island. The remains of this cat show that although it may well have been a valued human companion it was still as large as a wild cat belonging to the *libyca* race.

Right Mother comparing her son,
who is holding a cat, to the legendary
'Golden Boy' Kintoki, from the series
E-Kyodai by Kitagawa Utamaro
(1753–1806). Wood-block print.
(British Museum)

Far right Virgin and Child with a Cat
by Leonardo da Vinci (1452–1519).
Pen and ink over a sketch with a
stylus. (British Museum)

Above Footprint of a cat on a tile from Roman Silchester. (Reading Museum and Art Gallery)

Opposite Cat tethered to a chair leg, trying to detach the leash with a paw so that it can reach a bowl of food. Egyptian wall painting from the tomb of May (no. 130), Thebes, Eighteenth Dynasty, *c*. 1450 BC.

When a wild mammal is first bred in captivity and domesticated, the skeleton undergoes changes that are recognisable within a few generations, and strangely these changes appear to be very similar irrespective of whether the animal is a wolf, a pig or a cat. First of all the size of the body becomes much smaller, presumably as a result of an inadequate diet, stress and other factors related to its captive state. Most of the next changes to occur appear to be the result of the retention of juvenile characters into the adult state (sometimes called neotony). Life is very different for the tamed animal – its living conditions, its food, its daily rhythms and its reproductive activity become unbalanced. In response to the new way of life the focus of its perceptions is changed, and it never achieves the normal independence of a wild animal responsible for finding its own food and making its own nest or den. The tamed animal can be looked on as a perpetual child which is, of course, one of its functions as a companion to its owner. Hormonal changes occur, and perhaps in response to these the growth pattern is altered so that, although sexually and physically mature, the animal keeps some of its juvenile features. These are a short facial region to the skull, large round eyes, a reduced brain size, more body fat, a softer coat, and a more submissive personality. These changes can all be seen in the present-day tabby cat if it is compared with the wild cat of either the European or the African race. It is very difficult, however, to trace the process of domestication in the fragmentary remains of cats that are found on archaeological sites: a small piece of skull or a leg bone is unlikely to carry the criteria that will prove the animal to have been a household cat.

The evidence for domestication has therefore to be looked for in the archaeological context. For example, in the excavations of the Roman town at Silchester in England footmarks of a cat were found on tiles and, as it is unlikely that a wild cat would be walking about a brickworks where tiles were being made, it can be assumed that the Romans had domestic cats at Silchester.

It is generally assumed that the cat was first domesticated by the ancient Egyptians, but if this is so it was remarkably late, and long after all the other major domesticates had become fully established. There are no remains of cats from the prehistoric period in Egypt nor from the Old Kingdom (2686–2181 BC). From the Middle Kingdom there are pictorial representations of

cats, such as the beautiful picture of a cat in a papyrus swamp
from the tomb of Nebamun (Eighteenth Dynasty, *c.* 1400 BC),
but this could be a wild animal. The domestic cat appears only
from the New Kingdom onwards (from 1567 BC), and there are
many well-known pictures of cats from this period sitting under
chairs, hunting birds, and associated with other tame animals

Left Bronze Egyptian cat, the living form of the goddess Bastet. A votive figure dedicated by a wealthy worshipper of Bastet during the Saite period, *c*. 600 BC. (Presented to the British Museum by John Gayer-Anderson (Pasha) and Mary Stout, 1939)

Opposite Bastet, the Egyptian cat-goddess, with kittens, 4th century BC. She is seen here holding a *sistrum* (wire-rattle used in Isis worship) and an *aegis* (shield). (British Museum)

Below Egyptian cat, of glazed earthenware, Roman period, 1st century AD. It was most likely a good luck charm from the cult centre at Bubastis of the cat-goddess Bastet. (British Museum)

[32]

such as monkeys. It was from this time that the cat became a sacred animal in Egypt, but it must have been a common household animal long before then.

It may be that the cat first developed a symbiotic relationship with humans in very early times, the partnership being based around the rats and mice that would have been a pest in the home and in all stores of food. As today on farms, the cat would have obtained a plentiful supply of its natural prey, and the humans would not only have had their pests removed but would also have had a fireside companion.

Until they diffused northwards with human travellers rats and house mice were restricted to those regions where they were indigenous as wild animals and it is here, it would be expected, the cat was first domesticated. The black rat (*Rattus rattus*) was a native of Asia Minor and the Orient, the brown rat (*Rattus norvegicus*) spread from eastern Asia but did not reach northern Europe until the late Middle Ages, and the house mouse (*Mus musculus*), which now has a worldwide distribution, may have been restricted to southern Europe and Asia. No finds have been made of house mice on archaeological sites in Britain before the Iron Age (*c.* 500 BC) and no finds of rats before the Roman period. Rütimeyer, writing in 1866 about the animal remains from the Swiss Neolithic lake villages (dating from 3000 BC), had this to say on the question: 'The little animals which are so inconvenient in our modern houses, rats and mice, do not seem to have troubled the inhabitants of the lake dwellings, so that they could the more easily do without cats, of which no remains whatever are found.'

Zeuner in his classic work *A History of Domesticated Animals* (1963) discussed the origins of the names that are used for the cat, which he claimed came from the Near East and Africa, thus giving further support to the view that it was in this region that the cat was first domesticated. He quoted the theory that the word 'cat' is of North African origin and comes from the Berber languages. 'Puss' is believed to be derived from the name of the Egyptian goddess 'Pasht', Bastet or Bubastis. 'Tabby' may be of Turkish origin coming from 'utabi', although another theory holds that the name is derived from the name of a striped cloth first manufactured in the twelfth century AD and called after Attab, a prince in Baghdad.

Cat herding geese from the Satirical Papyrus, Egypt, New Kingdom *c*. 1500–1200 BC. (British Museum)

ANCIENT EGYPT

Whatever their early origins, it is clear that by the second millennium BC the cat was a fully domesticated animal in ancient Egypt, and it must have been descended from the local race of the wild cat, *Felis silvestris libyca*, for the paintings of the cats are extremely similar to the present-day wild form.

At a later period in Egypt all wild animals were made sacred, as described by Herodotus writing in about 450 BC:

Egypt, though it borders on Libya, is not a region abounding in wild animals. The animals that do exist in the country, whether domesticated or otherwise, are all regarded as sacred. If I were to explain why they are consecrated to the several gods, I should be led to speak of religious matters, which I particularly shrink from mentioning . . . . When a man has killed one of the sacred animals if he did it with malice prepense, he is punished with death, if unwittingly, he has to pay such a fine as the priests choose to impose (Bk II, Ch. 65).

In the next section (Ch. 66) the habits of cats are described:

On every occasion of a fire in Egypt the strangest prodigy occurs with the cats. The inhabitants allow the fire to rage as it pleases, while they stand about at intervals and watch these animals, which, slipping by the men or else leaping over them, rush headlong into the flames. When this happens, the Egyptians are in deep affliction. If a cat dies in a private house by a natural death, all the inmates of the house shave their eyebrows.

(Ch. 67) The cats on their decease are taken to the city of Bubastis, where they are embalmed, after which they are buried in certain sacred repositories.

The number of cats that were embalmed in ancient Egypt is quite extraordinary, and it was not only cats: all animals appear to have been mummified and wrapped in linen bandages, sometimes decorated, as were the human corpses. With the animals it is noteworthy that the most commonly retrieved mummies are those whose prey were rats, mice, snakes and other household pests. The most abundant animal mummies are cats, mongooses, ibises, vultures, hawks and crocodiles.

At the end of the last century so many mummified cats were excavated from Bubastis and other sites that boatloads of them were brought to Europe. There was then a glut on the market, and consideration was given to the idea of using them as fertilisers to spread on fields and as ballast for ships. Almost every small museum received its statutory mummified cat, but unfortunately, as so often happens when there appears to be a surfeit of some commodity, the finite end to the supply was not appreciated. Out of one consignment of nineteen tons of mummified cats imported to England, only one skull was kept and this is now in the Natural History Museum.

Besides the one skull from this collection there is a series of 190 skulls of cats from Gizah that was presented by the famous Egyptologist Flinders Petrie in 1907. Three of these skulls are much larger than the rest and can be clearly identified as belonging to the jungle or marsh cat, *Felis chaus*. This is a wild species which is found from North Africa through the Middle East to India. Like the mongoose the marsh cat may have been kept in captivity by the ancient Egyptians but it is very unlikely to have contributed to the genetic constitution of the domestic cat. The other 187 skulls can be easily recognised as belonging to *Felis silvestris libyca*.

Apart from these skulls, which have been removed from their mummy wrappings, there is in the Natural History Museum a

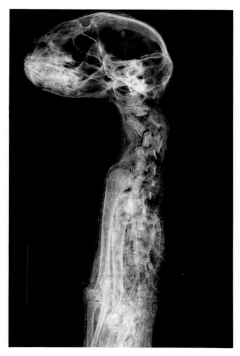

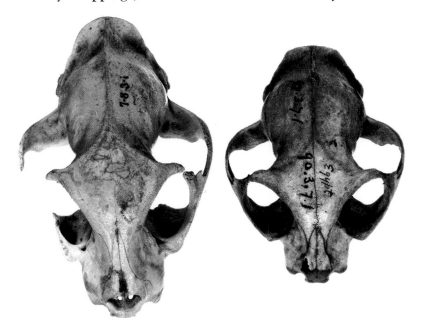

Above X-ray of mummified cat to show vertebrae displaced as a result of 'breaking the neck'. (Natural History Museum)

Left Skulls from mummified cats: *left Felis chaus; right* the one skull remaining from the nineteen tons of cat mummies imported to England. (Natural History Museum)

Far left Mummy of an Egyptian cat from Abydos, Roman period, after 30 BC. (British Museum)

Jungle or marsh cat, *Felis chaus*. From J. Anderson, *Zoology of Egypt; Mammalia*, London, 1902.

collection of fifty-five cats that are still wrapped and were also presented by Flinders Petrie. In a scientific study these mummies were X-rayed, and this revealed a number of very interesting features about them. After the animal had died its guts were removed and the body cavity was often filled with earth or sand. Then, in order to produce a compact mummy, the head was set at right angles to the neck, the forelimbs were stretched down along the front of the body, while the hindlegs were tucked up against the pelvis, and the tail curled up between the feet. When the body was still fresh, it was wrapped in bandages that had been soaked in natron (sodium salts) or treated with resin, and in some examples the wrappings were made into an elaborate pattern as with human mummies of the Ptolemaic period. A radiocarbon date carried out on one of the cats showed that it did, indeed, belong to this period which stretched from 330 to 30 BC, at least a hundred years after the writings of Herodotus (Armitage and Clutton-Brock, 1981).

Perhaps the most interesting point to emerge from the study of this collection of mummified cats was that contrary to the general belief that the ancient Egyptians never killed their cats many of these individuals had 'broken necks'. This could be seen in the X-rays as markedly displaced vertebrae in the neck. Another point was that in those mummies that could be aged the cats fell into two groups: twenty of the cats were between only one and four months old when they died or were killed, and seventeen cats were between nine and seventeen months; only two cats lived to be more than two years old. So it looks as though the cats were being specially bred in order to be mummified by the priests who perhaps sold them to the populace as votive offerings, which were then placed in sacred repositories. No doubt the value of the mummy depended on the degree of care that went into its preparation.

Sometimes the makers of the mummies cheated, the remains of different animals, or animals and human, being combined in the same wrapping. For example, a mummy has recently been investigated by Wolfgang Pahl which had the outward appearance of being a cat but on X-ray was found to consist of a cat's skull placed on top of four fragments of human tibia and fibula. This mummy, now in the Niagara Falls Museum, Ontario, is of Ptolemaic date (Pahl, 1986).

THE CLASSICAL WORLD

By the first millennium BC the cat was beginning to spread around Europe and Asia as a domestic animal. It is probable that during this early period there was a certain amount of inter-breeding with local races of wild *Felis silvestris* which would have been a much more common member of the wild fauna than it is today. The tabby cat of the north is much more like the Scottish wild cat in appearance than the African wild cat. There could be two reasons for this: the early domestic cats would have been as subject to environmental selection as the wild form and would have responded by becoming more heavy-bodied and thicker-furred, and would have developed shorter ears and tails. This selection towards the *F. s. silvestris* type could have been exacerbated with interbreeding and introgression of the wild genes.

There have been few new finds of early domestic cats since Zeuner wrote his classic work *A History of Domesticated Animals* in 1963, except for the important remains of a cat in the prehistoric levels of Cyprus (see p.26). Zeuner figured an ivory statuette of a resting cat from the site of Lachish in Israel, dated to around 1700 BC, and he mentions a terracotta head that has been identified as a cat from the Minoan site of Palaikastro on

Above Marble relief from Popopoulos, near Athens, with two people encouraging a fight between a dog and a cat, *c*. 510–500 BC. (National Archaeological Museum, Athens)

Below Ivory statuette of a resting cat from Lachish, Israel, *c*. 1700 BC. (British Museum)

[39]

Above Mosaic of a cat seizing a bird, Rome, 2nd century AD. A very similar mosaic was found at Pompeii.

Right Greek vase painting with two figures, Peitho and possibly Aphrodite (seated), who holds a pigeon at which a cat springs. 6th century BC. (British Museum)

Far right Head of a cat from the 'Lindisfarne Gospels', *c.* AD 700. (British Library)

the eastern shore of Crete (*c.* 1400 BC). These early records are rare finds, but from about 500 BC the cat is increasingly found in artistic representations and as skeletal remains at archaeological sites. The most famous early depiction is perhaps the marble relief from Popopoulos, near Athens, which shows two people who appear to be encouraging a fight between a dog and a cat.

Zeuner claimed that it was the coming of Christianity to Egypt that released the restrictions on the movement of sacred animals and allowed the barter and exchange of cats with the Romans. It is at least as likely that the spread of cats throughout the Classical world was an accompaniment of the spread of the brown rat and the house mouse. Most references to cats in the Roman literature are in the context of Egypt, but Pliny (Bk 10, Ch. 94) makes this interesting comment: 'Cats too, with what silent stealthiness, with what light steps do they creep towards a bird! How slily they will sit and watch, and then dart out upon a mouse!' Pliny wrote this in his *Natural History*, two years before he died in AD 79. The cat in the Classical world was probably valued more as a pest destroyer than as a fireside pet, and it never achieved the popularity or status of the dog.

For whatever reason cats spread rapidly throughout Europe in the Roman period. In Central Europe Bökönyi (1974) has reported the finding of forty-nine bones, representing fourteen individual cats, from the Roman villa settlement of Tac in Hungary. In Britain a nearly complete skeleton of a cat that died in a fire was retrieved from the Roman villa at Lullingstone in Kent; another was exavated from under a tessellated floor of a villa at Latimer in the Chiltern Hills, Buckinghamshire.

THE MIDDLE AGES

Following the end of the Roman Empire, Bökönyi states that there have been very few remains of domestic cats found on archaeological sites of the Migration period (up to AD 800) in Central Europe. In Britain, however, cat remains have been excavated from a high proportion of Saxon sites, of comparable date. At the site of Sandtun in Kent a skull and nearly complete skeleton of an adult domestic cat were retrieved, and from Saxon Thetford the remains of at least eighteen cats and kittens have been identified. By the Saxon period it seems clear that the domestic cat was as familiar an animal as it is today, at least in Britain, although obviously its numbers were still fairly low.

In the wonderfully illustrated 'Lindisfarne Gospels' produced about AD 700 there is a picture of a head of a cat and in the 'Book of Kells', written about a hundred years later, there are many pictures of cats, portrayed with great freedom of style, along with other domestic animals.

The value of the cat is given in the famous Laws of Hywel Dda, the Welsh king. These Laws are thought to have been written around AD 945, or somewhat later, and were translated into modern English in the nineteenth century. They have this to say about the cat: 'The price of a cat is four pence. Her qualities are to see, to hear, to kill mice, to have her claws whole, and to nurse and not devour her kittens. If she be deficient in any one of these qualities, one third of her price must be returned.'

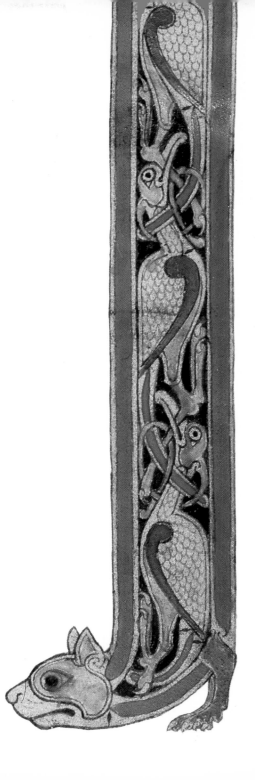

The Laws are very detailed and provide a fascinating picture of the relative values, not only of animals in the home and on the farm, but also of all kinds of household goods and farm equipment. The cat at four pence had the same value as the dog of a foreigner and a common house dog. This was more than the value of a little pig, a lamb, or a goose, which was one penny, but a great deal less than the King's greyhound which was worth 120 pence, or a hawk's nest which was worth one pound.

In another section of the Laws (the Gwentian Code) there is the following stipulation: 'Whoever shall catch a cat mousing in his flax garden, let its owner pay for its damage.' Flax was, of course, grown for the production of linen, a very valuable resource before the import of cotton into Britain.

It is of interest that there is no mention in the Laws of Hywel Dda of a value for the skin of the wild cat, although the skins of the fox (eight pence), the wolf (eight pence), the otter (eight pence), and the marten (twenty-four pence) are all given. The beaver, in the tenth century, was almost extinct in Wales, and indeed in the British Isles, and its value was set at 120 pence.

The wild cat was later to be included as one of the beasts of venery, and there is this enigmatic statement about it in 'The Nine Huntings' or 'The Hunting Laws of Cambria', translated with the Laws of Hywel Dda by Probert in 1823:

By the term squirrel is to be understood every animal that climbs a tree in its own defence. Therefore the huntsman ought not to say, marten, wild cat, squirrel, or fitchet, but to call them the grey climber, the black climber, the red climber. And because a climber is not able to run far, but ascends a tree, it is hunted by baiting and barking.

In the Pyrenees, and probably throughout Europe, where they occurred, both the lynx and the wild cat were hunted by royalty and they are described and illustrated in the famous book, *Livre de la Chasse*, by Gaston Phoebus, Comte de Foix, written before 1391, the year of his death. Between 1406 and 1413 this magnificent work was translated into English by Edward, second Duke of York, and he added to the text descriptions of hunting in England. Of the wild cat he had this to say (taken from the transcription of Baillie-Grohman, 1904): 'Of common wild cats I need not speak much, for every hunter in England knows them, and their falseness and malice are well known. But

Above Cat and mouse from the Luttrell Psalter, *c*. 1330. (British Library)

Opposite Two intertwined cats from the 'Book of Kells', *c*. AD 800. (Trinity College Library, Dublin)

one thing I dare well say that if any beast has the devil's spirit in him without doubt it is the cat, both the wild and the tame.'

It was at this time that the cat, and in particular black cats, were beginning to be considered as symbols of Satan and associated with witchcraft (see p.55). The wild cat remained, however, as a beast of venery, and therefore it was given a measure of royal protection. Baillie-Grohman quotes a number of royal deeds which granted licences to kill the beasts of the chase in forests throughout Britain. As early as 1205 Gerard Camoile had a special licence to hunt the hare, fox and wild cat throughout all the King's forests. Three hundred years later, in 1528, the Earl of Atholl entertained James V with a great hunt which lasted three days. Hart, hind and other small beasts such as roe and roebuck, wolf, fox and wild cats were killed.

The skins of cats were certainly used for the trimming of clothes in the early medieval period, but they may not have been highly valued. Again according to Baillie-Grohman, in the year 1127, in the Canons of Archbishop Corbeuil, it was decreed that 'no abbess or nun use more costly apparel than such as is made of lambs or cats skin'.

Perhaps the best description of a cat written during medieval times was by Bartholomew who produced his great encyclopedia of nature, *De proprietatibus rerum*, possibly as early as 1230. Bartholomew was English, but the book was written while he was in Paris. It was translated from the Latin into English by John de Trevisa, chaplain to Sir Thomas Berkeley, in 1397, and it remained the standard work in English on natural history until Shakespeare's time. Bartholomew's description of the cat is quoted here from *Animals in Art and Thought to the End of the Middle Ages* by Klingender (1971, p. 358):

He is a full lecherous beast in youth, swift, pliant and merry . . . and is led by a straw and playeth therewith: and is a right heavy beast in age and full sleepy, and lyeth slyly in wait for mice . . . In time of love is hard fighting for wives, and one rendeth the other grievously with biting and with claws. And he maketh a rueful noise and ghastful, when one proffereth to fight another; and unneth is hurt when he is thrown off an high place.

A later and better-known description of the cat, written in the form of a short essay, was produced by Edward Topsell in his

Above Wild cats being hunted by dogs and men with spears, from the *Livre de la Chasse* by Gaston Phoebus, Comte de Foix, before 1391.

Opposite A striped tabby and a tortoiseshell cat meet face to face.

Above right Cat playing a tabor and ass playing a trumpet, from Queen Mary's Psalter, early 14th century. (British Library)

Opposite Three cats and a rat from the 'Harleian Bestiary', 13th century. (British Library)

Below Cat from *The Historie of Foure-Footed Beasts* by Edward Topsell, 1607.

work *The History of Four-Footed Beasts* in 1607. This book was in the main a translation of the famous Swiss naturalist Konrad Gesner's five-volume work, *Historia Animalium*, written between 1551 and 1587. Like Bartholomew, Gesner and Topsell attempted to separate fact from legend in their descriptions of animals, but as with all other writers on the living world in the Middle Ages they were still restricted in their authorities to the Classical writers such as Aristotle, Xenophon and Pliny. Topsell's descriptions of animals, of which the cat is one of the most detailed, are some of the latest to be produced before the beginning of the eighteenth century and the sudden explosion of interest in the animal world which resulted in the great works of Buffon, Pennant, Bewick, and other naturalists and artists.

Topsell's account of the cat is particularly noteworthy for its mention of the allergy that humans can have to cat hair. He comments: 'It is most certain that the breath and savor of cats consume the radical humor and destroy the lungs, and they who keep their cats with them in their beds have the air corrupted and fall into hectics and consumptions.' And further on: 'The hair of a cat being eaten unawares stops the artery and causes suffocation.'

[47]

The source of most paintings of animals in medieval times is to be found in the bestiaries. These books became popular in the second half of the twelfth century and they can be associated with the transition from the Romanesque to the Gothic style of architecture. The bestiaries were religious books written in the form of animal parables and lavishly illustrated with all kinds of animals, both real and legendary. In the 'Ashmole Bestiary' there is a delightful picture of a cat licking itself.

A further source of medieval and later animal art is in the carvings to be found in churches. At Old Cleeve in Somerset there is a fifteenth-century tomb which has the figure of a man on it lying with his feet on a cat which, in turn, has its paws resting on a mouse. Another example can be seen in Winchester Cathedral where there is a cat and a mouse carved, *c*. 1305, on a misericord.

Above Carving of a cat with a mouse in its mouth on a misericord in Winchester Cathedral, *c*. 1305.

Opposite Three cats, one with upthrust hind leg while performing its toilet, from the 'Ashmole Bestiary', late 12th to early 13th century. (Bodleian Library, Oxford)

The Cat in Legend and Witchcraft

'A cat heard that there were some sick hens on a farm. So he disguised himself as a doctor and presented himself there, complete with a bag of professional instruments. Outside the farmhouse he stood and called to the hens to ask how they were. "Fine", came the reply – "if you will get off the premises."' The moral is: A villain, try as he may to act the honest man, cannot fool a man of sense (Handford, 1954). If it really originates from Aesop and has been translated correctly from the ancient Greek, this fable has a double interest for it indicates that both the chicken (an import from India and the East) and the cat were already familiar domestic animals in Greece about 550 BC, when Aesop is believed to have told his fables. Two other fables by Aesop concerned with a cat and mice show that the function of the cat as a pest destroyer was well known in Greece in the middle of the first millennium BC. These are probably the earliest stories known about cats from the western world.

During the Christian era many legends concerning cats have been passed down in folklore, but since medieval times most have represented the cat as a witch's 'familiar' or as a disciple of the devil. The earliest English legend, which appears to put the cat in a good light, is that of Dick Whittington and his cat. Dick was a poor boy who came to London with nothing but his cat; he made his fortune, and in the end was thrice Lord Mayor of London. This legend may date back as early as the late 1300s, but it has been suggested that the 'cat' was not the feline animal but the kind of heavy ship known as a 'cat' which was used for carrying coal from Newcastle to London. After this time the cat

Above Engraving of Richard Whittington, Lord Mayor of London, with his cat.

Opposite Cat receiving a deputation of mice. Engraving by Wenceslaus Hollar (1607–77) from *Aesop's Fables*, an edition by John Ellis, 18th century. The moral is: 'Trust not too rashly to a face as index of an inward grace.' (British Library)

[51]

Above 'Belling the Cat', from the painting of *Netherlandish Proverbs* (1559) by Pieter Brueghel the Elder. (Staatliche Museen, Berlin)

Opposite Three witches with their cats on their laps. 19th-century print.

was always seen as a villain, and in fairy stories every witch has a black cat, although oddly enough black cats are supposed to bring their owners good luck.

In the prologue to William Langland's *Vision of Piers Plowman* written in 1362–3 there is a description of 'belling the cat' which became a favourite topic for the illuminators of books and the carvers of choir stalls (according to Klingender, 1971). In the allegory a certain rat suggests fixing bells and a collar round the neck of a cat which has been plaguing the rats and mice. The bells are brought but no one dares to fix them on the cat. Then a mouse speaks up to say that even if they get rid of this cat another one, or its kitten, will come to persecute them in its place. By means of these symbols Langland made this passage understandable to everyone. The cat was King Edward III, the kitten who might replace him was his grandson Richard, then heir to the throne, the rats and mice were the commoners. The fable became well known and was narrated in a speech by Lord Gray to the conspirators against the favourites of King James III, at which Archibald, Earl of Angus, exclaimed, 'I am he who will bell the cat'; from which occasion he was to become known as 'Archibald Bell-the-cat'.

there were gossips sitting there,
By one, by two, by three:
Two were an old ill-favour'd pair:
But the third was young, & passing fair,
With laughing eyes, & with coal-black hair;
A daintie quean was she!
Rob would have given his ears to sip
But a single salute from her cherry lip.

As they sat in that old and haunted room,
In each one's hand was a huge birch broom,
On each one's head was a steeple-crown'd hat,
On each one's knee was a coal-black cat;
Each had a kirtle of Lincoln green —
It was, I trow, a fearsome scene.

Now riddle me, riddle me right, Madge Gray,
What foot unhallow'd wends this way?
Goody Price, Goody Price, now areed me right,

Who roams the old Ruins this drearysome night?'
Then up and spake that sonsie quean,
And she spake both loud and clear:
'Oh, be it for weal, or be it for woe,
Enter friend, or enter foe,
Rob Gilpin is welcome here! —

'Now tread we a measure! a hall! a hall!
Now tread we a measure,' quoth she —
The heart of Robin
Beat thick and throbbing —
'Roving Bob, tread a measure with me!'
'Ay, lassie!' quoth Rob, as her hand he gripes,
'Though Satan himself were blowing the pipes!'

VI

From this time the cat gradually took on the character of evilness, of Satanism and of withcraft. The sixteenth and seventeenth centuries saw the terrible period of trials for witchcraft. Topsell, in 1607, wrote: 'The familiars of witches do most ordinarily appear in the shape of cats, which is an argument that the beast is dangerous to soul and body.' The first trial for witchcraft in England was in 1566, in the reign of Elizabeth I. Agnes Waterhouse and her daughter Joan were executed for being linked in witchcraft with a cat that was 'a whytte spotted catte... [and they] feed the sayde catte with breade and milkye... and call it by the name of Sathan'. The last official execution for witchcraft took place in 1684, and during this 118-year period the cat became progressively more feared as the witch's 'familiar', although other animals were also considered to carry evil spirits.

According to Tabor (1983), the belief that a cat has nine lives first arose from a statement in about 1560 in *Beware the Cat* by Baldwin who wrote, 'it was permitted for a witch to take her cattes body nine times'.

Children's literature provides an abundant source of folk tales and rhymes on cats, as on many other animals, and the works of Iona and Peter Opie may be consulted for reference to these (1951, 1959, 1969). Many of the rhymes are examples of folk memory that go back to the time when cats were persecuted.

Opposite Puss in Boots by Walter Crane (1845–1915). Pen, brush and Indian ink. (British Museum)

Below Woodcut of three witches and their cats from a 17th-century broadsheet reporting their trial.

Throughout the sixteenth and seventeenth centuries cats were subjected to appalling torments in the cause of searching out the devil, especially during Lent when it was customary to throw them into bonfires and so on. Later, during the eighteenth century, cats were very cruelly treated in all sorts of baiting 'sports'.

One superstition of which there is material evidence was that if the body of a cat, or better still the bodies of a cat and a rat, were built into a house wall they would keep away rats. This belief survived as late as the eighteenth century, and the dried mummified corpses of cats have been found in a fairly large number of buildings in Britain and Europe. Some of these corpses may be the result of cats creeping into holes while a new house was being built and then being trapped. Most, however, have been dried in a life-like posture and then carefully built into the wall. A number of these dried cats are held in the research collections of the Natural History Museum, the finest being a cat with a rat placed just underneath it from a house that was recently demolished in Bloomsbury, London.

Left Dried corpses of a cat and a rat found in a house in Bloomsbury, London. (Natural History Museum)

Far left A cat found in Cheapside, London, 'suspended on a gallows, habited like a monk with shaven crown'. Engraving from *Fox's Book of Martyrs*, 16th century.

[57]

Cats in Modern Times

Most domestic cats throughout the world today still have coats that are variations on the tabby, either striped or blotched. Besides being nearest to the wild species in pattern, the striped tabby is also a coat that is typical of many wild mammalian carnivores and rodents. The pattern is called by biologists 'agouti', because it is exemplified by the South American rodent of this name. Each hair of the coat has a brindled grey, black and white appearance which is due to the uneven distribution of melanin (the dark pigment) throughout its length. The overall appearance of the coat is striped or banded with 'salt and pepper' flecks as a result of the distribution of the agouti hairs.

Although a large number of wild mammals have agouti coats, the pattern is uncommon in domestic animals apart from cats. The agouti pattern can be rather easily bred out through a non-agouti recessive gene which produces hairs that are uniformly pigmented along their length and may be of various colours such as red, black or white.

THE FOUNDATION BREEDS

When Darwin published *The Variation of Animals and Plants under Domestication* in 1868, he could find relatively little to write about the domestic cat. He noted that the cats of India and South America were different in appearance from those in Europe, but his main observation was that the nocturnal and rambling habits of cats prevented artificial selection for breeds. Apart from variation in coat colour, the only breeds that Darwin

Above Blotched tabby balancing adroitly on a garden fence.

Opposite Striped tabby.

mentions in Britain are the Manx and the Persian, about which he writes: 'The large Angora or Persian cat is the most distinct in structure and habits of all the domestic breeds; and is believed by Pallas, but on no distinct evidence, to be descended from the *Felis manul* of middle Asia.'

One aspect of the complicated relationship between people and their domestic animals is concerned with the desire for ownership of the rare, the beautiful and the exotic. Kings in the ancient world kept cheetahs, lions and many other strange and wonderful creatures to enhance their status, and people every-where, and at all times, have enjoyed owning an animal that is 'special'. Keith Thomas in his fascinating book *Man and the Natural World* (1983) comments that 'Archbishop Laud was particularly fond of cats and in the late 1630s was given one of the earliest imported tabbies, then valued at £5 each, but soon to become so common as to supersede the old English cat, which was blue and white'. These 'new' tabbies were probably the blotched type which it has been postulated first arose as a new form in Elizabethan times, although it is interesting that the famous illustration to Topsell's account of the cat (published in 1607) is still of the striped tabby.

The selective breeding of cats to produce new and distinctive forms began during the second half of the nineteenth century at a time when there was a very wide-scale interest in the 'improvement' of all domestic animals. This was a result of the new understanding of the process of evolution following the publication of Darwin's works, but it was also the result of the industrial revolution, urbanisation and the increased leisure of people who worked in the manufacturing industries. There was a quite new fascination with the keeping of animals as pets, from cage birds to cats, and many exotic breeds were imported such as Pekingese dogs and Siamese cats.

The breeding of cats achieved greater popularity in Britain than anywhere else, and the first cat show was held on 16 July 1871 at the Crystal Palace. It was organised by Harrison Weir who had a great interest in promoting the welfare of cats as well as in producing and defining new breeds.

From the first there were two main types of breeds: the British (or European or American depending on the country of origin) which is the cold climate form with stocky body, large

Above Brown tabby from *Our Cats* by Harrison Weir, 1889.

Opposite L'hiver: chat sur un coussin, 1909. Colour lithograph by Théophile Steinlen (1859–1923). (British Museum)

Below Spotted tabby from *Our Cats* by Harrison Weir, 1889.

Above Manx.

Opposite Black short-haired cat with typical cold climate body shape.

head, rather short ears and thick coat. The second is the Foreign which has the hot climate form of slender body, long limbs, large ears and shorter coat.

If these two types are compared with the wild ancestor, it can be seen that the British breeds look like the Scottish wild cat, and the foreign breeds look like the African wild cat (see p. 9). These similarities will have evolved by natural selection in response to environmental conditions acting on the domestic animals but also as a result of some interbreeding with the wild forms.

One of the earliest Foreign breeds to be brought into Britain was the Siamese, and individuals of the breed were shown to the public at the Crystal Palace in 1886. They caused a sensation with their remarkable coloration and slender bodies, and since then all Foreign breeds have had the same body shape.

Within the types of body shape another classification of breeds is based on the length of coat, with the Persian being the exemplar of the long-haired breeds. This has the British type of body, although the breed did, most probably, originally come from western Asia. It is likely, however, that interbreeding with English cats at the end of the last century, which kept the long hair, at the same time altered the body and head shape towards a more stocky conformation.

Besides the indigenous domestic cats which may have been introduced by the Romans or even earlier, there were at the end of the last century in Britain the following exotic breeds whose histories will be summarised briefly:

Manx
It is possible that the mutation causing the absence of a tail arose independently on the Isle of Man and that cats with no tails were selected until the breed became established. On the other hand, stubby-tailed cats are fairly common in eastern Asia. It is the result of a dominant gene that can cause a complete lack of the tail vertebrae or produce a short or stumpy tail.

There is a legend, first exposed in print at the end of the last century, that the Manx cat is descended from cats that swam ashore from a sinking ship of the Spanish Armada in 1558. This explanation for peculiar local breeds of domestic animals has been applied to a variety of other species including Connemara

Above Royal cat of Siam, from *Our Cats* by Harrison Weir, 1889.

Opposite Siamese mother cat licking her kittens.

ponies and the four-horned Jacob's sheep, and it is hard to know whether it could have any foundation in fact. Perhaps it is more likely that these distinctive breeds are the result of artificial selection of a small gene pool in an isolated environment.

Siamese

This undoubtedly ancient breed first reached England towards the end of the nineteenth century. It is claimed that Siamese cats were originally kept only by members of the Siamese royal family, and it is recorded that it took years of negotiation by the daughter of General Walker before one male and two females were allowed to be exported to England.

The original Siamese cats were of the seal-point variety and they often had squinting eyes and kinked tails. Kinky tails are a characteristic that is often seen in common cats in eastern Asia, but both this and the squint have been bred out of most Siamese cats of the present day.

A feature of the coat of the Siamese is that a warm temperature tends to cause the coat to grow light in colour. Siamese kittens are born quite light all over, and the points then become darker with age. The body is that of a typical hot-climate felid with a long head, large ears, a svelte body and a fine tapering tail.

The Siamese cat differs in a number of characteristics from European cats. Besides its remarkable coat colour it has a gestation period on average five days longer than other cats, and the size of the litter is larger than average. The voice is very different, the temperament is excitable, and the father is claimed to be more friendly to the kittens than is usual amongst other cats. In general the Siamese is a more social animal than other domestic cats. All these characteristics were investigated by Hemmer in a recent study into the origins of the domestic cat and the possibility of their evolution from species other than *Felis silvestris*. Hemmer compared the behaviour, litter size, gestation length and vocalisations of a pair of captive African wild cats (*Felis silvestris libyca*) with those of Siamese and Persian cats, and he also examined the relative sizes of the cranial capacities of the skulls of all the different types of cat. He concluded from this analysis that there was no evidence to indicate that either the Siamese or the Persian cat was de-

Siamese cats by Christopher Wood
(1901–30). Coloured chalks, 1927.
(British Museum)

scended from any other species. The peculiar vocalisation of the
Siamese fell within the range of the wild cat, and the social
behaviour was also consistent with that of the pair of wild cats
where the female gave birth with the male present in the same
cage; he licked the kittens and generally showed no aggression
towards them. Hemmer found that long gestation was correlated
with large litter size in the Siamese.

Besides the original variety of Siamese which is known as seal-
point there are, today, the chocolate-point, the blue-point and
the lilac-point varieties. All of these have the typical form of the
Siamese and all have blue eyes, but the shading of the body and
points varies. Other point-colours are classed Colourpoint
Shorthair.

Abyssinian

Of all breeds the Abyssinian is closest in appearance to the presumed ancestor of all domestic cats, the African wild cat, *Felis silvestris libyca*. It must not be assumed, however, that this is because the Abyssinian is a direct descendant of the African cat or of the ancient Egyptian cats that are so well known from tomb paintings and their mummified bodies.

There are two conflicting accounts of the origin of this breed, but the facts are still unknown. Dr Gordon Staples wrote in 1882 that a Mrs Barrett Lennard brought a cat of this type back with her from Abyssinia (Ethiopia) at the end of the Abyssinian war. On the other hand, Harrison Weir, the instigator of the first cat show, believed that the Abyssinian was the result of intentional breeding in Britain from tabby stock. Pocock, writing in 1907 on English cats, was unable to decide on their origin. He believed that the breed could either have come from Abyssinia or it could be derived from what he described as the 'ticked' type of striped tabby. In either case it seems certain that the Abyssinian must have been interbred with cats of foreign body type in order to produce the long head, lithe body and slender limbs.

Above Drawing by Théophile Steinlen (1859–1923).

Left Abyssinian.

[67]

Above Angora, from J.B. Huet, *Collection des mammifères du Muséum d'Histoire Naturelle*, Paris, 1808, pl. 17.

Below Persian.

Angora or Persian

Angora cats, as the Persian was first known, were the oldest long-haired variety of cat to be brought to England. They came from Paris and were known for some years as French cats. They were pure white with yellow or blue eyes (sometimes a mixture of both) and they had long silky hair and short faces. There is a genetic link between colour of the coat and eyes and deafness so that white cats, particularly those with blue eyes, have a tendency to be deaf.

The earliest white cats may have come from the mountainous region near Lake Van in Turkey where white cats are common today. Some years later white cats with long hair were brought to Britain from Persia but they were rather different in appearance from the Angoras: their heads were not so long, the hair was not so soft, and the eyes were more often yellow. By 1903 apparently all the Angoras had been replaced by the Persian variety, probably because of the deafness problem which affected the blue-eyed cats most often. Although the first long-haired cats to appear in Britain were white, it was not long before other colours were added to the Persian breed, and there were black, red, smoky, blue and other coloured Persians at the early cat shows in the nineteenth century.

In his 1907 account of English cats Pocock noticed a feature of the Persian cat skull that has become progressively developed over the last eighty years. This is a marked shortening of the facial region of the skull with compaction of the teeth, an overshot lower jaw and great widening of the facial region. This shortening of the face has become so accentuated within recent years that the skull of the Persian cat bears a remarkable resemblance to that of the Pekingese dog, where the same effect was produced by hundreds of years of breeding by the Chinese.

It is often stated in the literature that the Angora and Persian cats are descended not from *Felis silvestris* but from Pallas's cat, *Felis manul*. This suggestion was first put forward by Pallas himself and quoted by Darwin in 1868 (see p.61), but it was strongly refuted by Pocock in 1907 on the grounds that the skull of Pallas's cat is quite different from that of the Angora or Persian (even before the modern, extreme modifications to the facial region).

Another suggestion has been that the Persian cat is descended

Above Cat and calligraphy, school of Shibata Zeshin (1807–91). Japanese hanging scroll, ink and colours on silk. The cat bears a remarkable resemblance to the Lake Van cat illustrated left. (British Museum)

Left Turkish Lake Van cat.

Right Cats from the *53 Post-stations of the Tokaido Road* – part of a triptych of various coloured cats in different attitudes by Utagawa Kuniyoshi (1797–1861).

Far right Close-up of a Siamese cat, with distinctive bright blue eyes.

Harrison Weir with the winner of the First Prize at the Crystal Palace Show – a Persian kitten.

however, of less importance than the production of a pure line of beautiful healthy cats that are a source of pride and contentment to their owners.

The incentive to breed and improve any domestic animal follows from the competition induced by 'the Show'. At the end of the last century the showing of animals came as a natural result from efforts to improve farm livestock and this was followed by shows of animals of less economic value: dogs, pigeons, cage birds and cats. It was Mr Harrison Weir, a Fellow of the Horticultural Society, an artist and a cat lover, who set up the first cat show, and who stated his motives as follows:

I conceived the idea that it would be well to hold Cat Shows, so that different breeds, colours, markings, etc. might be more carefully attended to, and the domestic cat sitting in front of the fire would then possess a beauty and an attractiveness to its owner unobserved and unknown because uncultivated heretofore.

In 1887 the National Cat Club was founded with Harrison Weir as its first President and a stud book was opened. The Club continued until 1910 when the Governing Council of the Cat Fancy was formed, but it continued to conform to the 'Points of Excellence' for breeding that had been laid down by Harrison Weir, and some of these standards still hold today. Harrison Weir devised a method of judging each cat that used a hundred-point system.

At first there were very few breed classes and so the cats were classified according to their colour: for example, there were classes for Black, White and Tabby Longhairs, and another large class for 'Any Other Colour'. From this small beginning there has grown today's boom in the breeding of pedigree cats with shows all over the world and the exchange of breeding stock between continents, so that for a large number of people the breeding and showing of cats is a way of life.

New breeds will continue to be developed but they are still based on the crossing and re-crossing of the long and shorthairs with the British and Foreign body shapes, and all derive from the breeds of the last century, these being primarily the striped and blotched tabbies, the single-coloured British types, the Abyssinian, the Siamese, the Persian, the Russian Blue, and the Manx or its Asian equivalent.

The main breeds of cat at the present day are:

Shorthair cats

British Shorthair	Oriental Spotted Tabby
American Wirehair	Korat
Exotic Shorthair	Egyptian Mau
Scottish Fold	Burmese and Malayan
Manx	Bombay
Japanese Bobtail	Russian Blue
Siamese	Abyssinian
Colourpoint Shorthair	Singapura
Havana Brown	Rex
Oriental Shorthair	Sphynx (Hairless)

Longhair cats

Persian/Longhair	Norwegian Forest Cat
Himalayan/Colourpoint Longhair	Birman
Turkish Angora	Balinese and Javanese
Somali	Maine Coon

Below Archangel Blue cat from *Our Cats* by Harrison Weir, 1889.

Overleaf (top) Sphinx (Hairless); *(bottom)* Burmese.

Page 77 (top) Egyptian Mau; *(bottom)* Oriental Spotted Tabby.

Dessins sans paroles

des Chats

par

Steinlen

PARIS
ERNEST FLAMMARION
ÉDITEUR

DISTRIBUTION OF THE COMMON CAT IN TOWN AND COUNTRY

The breeding of domestic cats and their wild progenitor, *Felis silvestris*, can be divided into four categories: first, the completely controlled breeding and selection that is carried out by the Cat Fancy; second, the uncontrolled breeding of common domestic cats where mating is indiscriminate but there is selection for preferred colour and sex in the kittens; third, the breeding of farm cats and feral cats (those that have returned to live in the wild) where there is no human control or selection; fourth, the breeding of the wild cat, *Felis silvestris*, away from human habitation.

Within recent years investigations into the worldwide distribution of the coat pattern, colour and length of hair of town and country cats in the second breeding category have led to some remarkable observations. The free-ranging household cat that mates without human interference is of ancient origin, and the lives of these cats have probably changed little in the last 2,000 years, except that those living in inner cities are, like their owners, under greater pressures from crowding and chance accidents. However, as humans have travelled around the world they have taken their cats with them, and there has been a cosmopolitan mixing of genetic stock as well as a greatly increased migration of the domestic cat into all habitable lands.

The remarkable fact that has been discovered about the distribution of cats is that their coat patterns and colours are not randomly dispersed amongst the different populations. Those in an urban environment differ from those in a rural area, and they may differ in the majority of cats between an inner city and its suburb. Furthermore, assessment of the most common type of coat and its colour in different cities and different countries can be correlated with the movements of people over the centuries, and it has even been used to make deductions about the approximate date when an immigrant group of cats, and hence their owners, arrived in a new region.

Most of the recent work on the distribution of coat colour in cats has been carried out by Neil B. Todd, Director of the Carnivore Genetics Research Center in Massachusetts, USA, and he has summarised his results in *Scientific American* (1977).

There are a number of coat colours that probably derive from

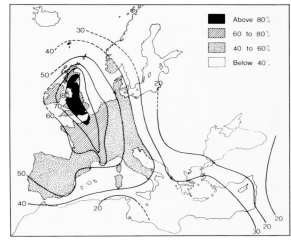

Above The density and distribution of blotched tabbies over Europe and western Asia. The figures are percentages. (After Todd, 1977)

Opposite Dessins sans paroles des Chats by Théophile Steinlen (1859–1923). (British Museum)
Below Reclining black cat.

Tortoiseshell cat on a rock watching butterflies under hollyhocks and other flowers, by Cai Han and Jin Yue (1647–86). Chinese painting. (British Museum)

mutations which occurred early on in the history of cat domestication because they have a worldwide distribution. They are tabby, black, blue, orange, white-spotted and white. The mutation for long hair may be combined with any of these colours and is probably also of ancient origin. To a certain extent the coat colour is self-perpetuating, but it is probably preserved as well by human selection for aesthetic reasons which include the important factor of rarity value, that is, ownership of a special animal; in this category must be placed the polydactylous cats. These are cats that have an extra digit or toe on some or all feet, with the extra toes being more common in some breeds and in some parts of the world than in others.

Although ubiquitous, the coat colours and long hair do not occur with equal frequency in all regions, and it is this uneven distribution which has been investigated by Todd and other geneticists.

The three most common coat colours throughout the world at the present day are the blotched tabby, black, and orange, which is sex-linked to produce ginger male cats and tortoiseshell female cats. In his study of the population genetics of cats Todd collated the numbers of cats of each of these three coat colours in a great many localities in Europe, western Asia, and North Africa and he then drew distributional (or clinal) maps from the results.

To take the black coat colour first, it was found that the highest frequencies were in Britain and north-west Africa with slightly lower numbers around the Mediterranean and in the Italian cities of Rome and Florence. Todd suggests that black cats originated in the eastern Mediterranean during the Classical period. It appears that the single-coloured coat, notably black, is favoured by the urban environment, perhaps because this mutation is linked through the hormonal system with a temperament that is less aggressive, less fearful and more tolerant of crowding than the wild type, that is, the agouti or striped tabby.

This placid temperament appears to be even more accentuated in the blotched tabby for which Todd also produced a clinal map. He deduced that the blotched tabby had a relatively recent British origin (since the time of Elizabeth I), as well as a less prominent second place of origin in north-eastern Iran.

Tortoiseshell cat.

Todd extended his sampling of the blotched tabby to America, Canada, Australia, New Zealand and Tasmania, which are all countries that had no indigenous cats until colonised from Europe over the last 300 years. He then suggested that the differences in the numbers of blotched tabbies may reflect the dates at which the various cities were first colonised: for example, Hobart and Dunedin, which are most similar to present-day England, may have been colonised by blotched tabbies later than Brisbane, with Adelaide having an intermediate position.

The distribution of orange cats is very different from that of the black and the blotched tabby. Besides being most common in the Orient, there are very high concentrations of ginger and tortoiseshell cats in Turkey, along the north coast of Africa, and

Above Seated ginger cat.

Opposite Striped tabby in tack room.

in the western islands of Scotland, with somewhat fewer in the Faroes, Iceland and the Isle of Man. Todd postulates a centre of origin for European orange cats in Asia Minor, and he suggests that the high numbers in the Western Isles may be a relic from Viking times. Whatever the true explanation, it does seem that orange cats have no particular survival value in large cities where they are never as common as black or blotched tabby cats.

The farm cat lives a very different life from the town cat. It has to be much more self-sufficient: indeed, many farm cats receive almost no food from humans and are expected to live entirely on rats, mice and scavenged food. They are obviously much less crowded in numbers than in the cities: Tabor (1983) gives the figure of one cat to twenty to twenty-five acres for rural England. The core area for the cat family will be the farm buildings, but the animals will probably range much more widely than the town cat, and their perceptual world is much closer to that of the wild cat. Like the populations of feral cats, which live quite outside human control, farm cats are subjected to very little artificial selection, and their survival will depend on how adept they are at fending for themselves. The placid blotched tabby is unlikely to do as well in this environment as the striped tabby, but no comparisons have yet been made of the relative numbers of different coat colours on the farm and in the city.

The surveys carried out by Todd and other workers on population genetics have examined a phase in the domestication of the cat that may be drawing to a close, at least in the industrial countries of the western world. This is because nowadays it is becoming increasingly common for male cats to be castrated and female cats to be spayed so their breeding is therefore no longer uncontrolled. People living in flats or small houses are unwilling to put up with nights interrupted by the fearful wailing of mating cats, and neither are they willing to put up with the smell of tom cats or the nuisance of kittens for whom they cannot find homes. In urban areas, at least, it may be that the natural selection that has promoted the single-coloured cat and the blotched tabby may be replaced by artificial selection for increased diversity of colour and breed. It will be interesting to see what the cats of the inner cities look like in, say, another twenty years time.

CATS IN THE ORIENT

The Siamese is the best-known breed from Asia in the western world. It is of ancient origin and, it is claimed, was restricted in ownership, until the end of the last century, to the Siamese royal family, but there are other breeds that are reputed to have an equally noble and ancient ancestry. The Birman, the sacred cat from Burma, may indeed be the ancestor of the Siamese. It looks like the Siamese but has white paws. Gebhardt in *A Standard Guide to Cat Breeds* (1979) recounts the following legend about the Birman:

Centuries ago in Burma white cats were guardians of the Temple of Lao-Tsun, which housed a golden goddess with deep blue eyes, Tsun-Kyan-Kse. One beautiful cat, Sinh, was the close companion of the head priest, Mun-Ha. As the priest and his cat sat in front of the goddess, the temple was attacked by raiders and Mun-Ha was killed while praying. Sinh placed his paws on his dying master and faced the golden goddess. As he did this, his white fur took on a golden hue. His yellow eyes became a deep blue, and his face, tail, and legs turned to the colour of the earth, but the part of his paws that touched the dead priest, remained white, the symbol of purity.

Birman, the sacred cat from Burma.

Opposite Japanese Bobtails.

Below Hollow earthenware cat painted with lustre details on a semi-opaque white glaze. Possibly from Rayy, Persia, 13th century. (British Museum)

The difficulty in believing in an ancient ancestry for the Birman is that long-haired cats are said to be very rare, if not unknown, in eastern Asia, probably because of the extremely hot climate, so it may be that this breed originated as a cross between the Siamese and the Angora or Persian. On the other hand, it is claimed that there were cats of the same coat pattern and coloration in the temples of Tibet where long-haired cats would be more explicable in the cold mountains.

Studies of the population genetics of common cats in eastern Asia have shown some marked differences in the distribution of coat pattern and colours from those in Europe. In 1959 Searle published a detailed account of the variations he observed in the cats of Singapore, and although written twenty years before the surveys of Todd, the comparisons are still of great interest.

Oriental cats are noted for two abnormalities in their tails which were described by Darwin in *The Variation of Animals and Plants under Domestication* (1868): 'Throughout an immense area, namely, the Malayan archipelago, Siam, Pegu, and Burmah, all the cats have truncated tails about half the proper length, often with a sort of knot at the end.' At the present day a large number of the cats of the East can still be seen to have this 'kinky tail', which is a feature never seen in European cats. In his survey of the Singapore cats Searle found it was present in 69.2 per cent but he did not find a single cat with a kinky tail in an earlier study he carried out on variation in London cats (1949). Searle also found that stubby-tailed cats were very common in Singapore but rare in London, and in general the stubby tail is rare in Europe, except in the Manx breed where the tail is more usually totally absent.

These abnormalities in the tails of Oriental cats could be interesting examples of what geneticists call a 'founder effect', and it could indicate that all the cats of South-east Asia are descended from a very few individuals that were brought into the region some hundreds or perhaps thousands of years ago. A founder effect is produced when a new population is derived from a limited number of individuals which represent a very small sample of the genetic pool to which they formerly belonged. Natural selection operating on this restricted variation then leads to gene combinations quite different from those of the ancestral population.

[84]

As stated above, long hair is very rare in cats from eastern
Asia, and Searle found no long-haired cats in Singapore and
none with extra digits, from the condition known as polydactyly.

The blotched tabby is seldom seen in the Orient, which
supports the hypothesis that this is a relatively new mutation
that has not yet spread worldwide. Searle found that single-
coloured cats, particularly black, were also rare compared with
London, but striped tabbies were common, supporting the
assumption that this is the foundation type of the domestic cat.
The Abyssinian type of coat with 'ticking' of the differently
coloured hairs, rather than stripes, was more common in
Singapore than in London, but most common of all was the
white-spotted coat and the orange, as they are throughout
eastern Asia.

FERAL CATS

There are cats that live wild all over the world: they can be found in Soho, or on Ascension Island, or in the deserts of Australia. These are not true wild cats like, for example, the Scottish wild cat because they are descended from cats that have been domesticated and have been later abandoned by their owners. Domestic cats that live in the wild are described as *feral* and they may continue as thriving breeding populations for hundreds of years. Alternatively, transitory groups of feral cats may intermingle with truly domestic cats so that an outsider may find it difficult to distinguish between a feral cat and a house cat, both scavenging for food in the same area, as every traveller to the Mediterranean will confirm.

In his book *The Wildlife of the Domestic Cat* (1983) Roger Tabor describes a number of studies that have been carried out in Britain on populations of feral cats. He claims that almost every colony of feral cats that he looked at in England was being fed by at least one person who sometimes spent a considerable amount of time and money on the task. Feral cats in English cities are also subject to a certain amount of control in that they are often trapped, castrated and then allowed to go free again.

Opposite Cat watching insects by Ran'ei, first half of 19th century. Japanese hanging scroll, ink and colours on silk. (British Museum)

Below Feral cats outside the city walls of an Italian town.

Many big companies, institutions and buildings have their own colonies of feral cats which have varying degrees of management, often depending on the wishes of the employees and their views on health and safety. Some companies agree to the maintenance of a number of cats that are neutered and identified by means of a collar. It may be noted that the Natural History Museum removed all its feral cats about twenty years ago, and at the British Museum members of staff formed a special committee – the Cats Welfare Society – to manage their cat colony. Houses were provided for the neutered cats on the flat roof of a low building and special feeding places were designated. The colony was finally dispersed about three years ago.

It is probably still true that feral cats in all cities throughout the world play a large part in the control of rodent pests, although nowadays there is so much readily available garbage that the cat does not have as much incentive to hunt as the farm cat where there is much less food to be scavenged.

In densely built-up areas like the centre of London colonies of feral cats can have strongholds that are difficult for the carefully nurtured house cat to penetrate. Tabor has described the complicated territorial contacts that are made between house cats and feral cats in this situation, which will also affect the breeding of the animals, while the owners know nothing about the problems faced by their pets.

In countries where the wild species of cat (*Felis silvestris*) is still found, a certain amount of interbreeding occurs between populations of feral cats and the wild cat. This is probably rare at the present day, at least in Europe, but in earlier times when the wild cat was still a relatively common wild carnivore it probably happened quite frequently, and it must in part explain the different body proportions of the north European, so-called British, and the southern or Foreign domestic cats (pp. 61–2).

Nowadays in Scotland hybrids between domestic or feral cats and the wild cat are sometimes discovered and they provide instant material for amazing speculations by the newspapers. Headlines such as this appear: 'Black Beast of the Highlands! Sheep killer is huge wild cat. Scientists have discovered that a black beast which has been killing sheep in the highlands is a hitherto undiscovered mutation of wild cat' (*Daily Express*, 15

December 1984). During the early 1980s a number of large black cats were killed on Speyside near the Cairngorms and were almost certainly the result of interbreeding between wild and domesticated or feral cats. They *were* larger than the normal domestic cat, but it was fascinating to watch the rapid evolution of the legendary, terrifying animal: first a large black cat, then a black cat as big as a dog, then a black panther roaming the countryside with the commentary:

Shot animal in 'panther' probe. Researchers to examine body of strange animal shot in area of Scotland where around 50 sightings of 'black panther' reported. Animal, jet black and as tall as dog, shot by gamekeeper near Dallas, Grampian as it approached pheasants. Blood sample sent to Aberdeen University (*Birmingham Evening Mail*, 16 October 1985).

Above Black cat from Speyside, most probably a hybrid between a wild and a domestic cat. (Natural History Museum)

Opposite (top) Susie, one of the British Museum's feral cats, who died in 1982; (*bottom*) feral cat patrolling the dustbins at Paddington, London.

Index

References in italic type are to pages with illustrations

Abyssinian 67, *67*, 74, 75, 85
Acinonyx see cheetah
Aesop *50*, 51
African wild cat 8, 9, *11*, 12, 13, 26, 30, 36, 37, 39, 62, 64, 67, 90
agouti 59, 80
Angora 68, *68; see also* Persian
Archibald, Earl of Angus 52
'Ashmole Bestiary' *48*, 49
Asia 12, 33, 39, 62, 80, 82, 84, 90
Attab 33

Balinese 75
Bastet *32*, 33, *33*
'belling the cat' 52, *52*
bestiaries *see* 'Ashmole Bestiary', 'Harleian Bestiary'
Birman 75, 83, 84, *84*
Bombay 75
'Book of Kells' 41, *42*
breeds 59, 61–2, 71, 74–5; British 61, 62, 70, 74, 75, 88; Foreign 62, 70, 74, 88; Longhair 74, 75; Shorthair 66, 74, 75
Bubastis 33, 36
Burmese 75, *76*

Cat Fancy 71, 74, 79
cheetah 6, 61
coat colours 25, 59, 71, 78, 79–80, 84; black 25, 45, 52, 68, 74, 79, 80, 82, 85; blue 68, 80; ginger 80, 81, 82; orange 80, 81, 82, 85; red 68; Siamese 64; smoky 68, 80; tortoiseshell *44*, 80, *80*, 81, *81*; white 68, 74, 80; white-spotted 80, 85
Colourpoint Shorthair 66, 75
Cornish Rex *see* curly-coated
Crystal Palace 61, 62, 70
curly-coated 71
Cyprus 26, 39

Darwin, Charles 59, 61, 68, 70, 84, 90

dog 6, 8, 17–18, 19, 21, 71
drop-eared 71

Edward III, King 52
Egypt 30, 33, 36, 38, 40, 67
Egyptian Mau 75, 77
European wild cat 6, 8, 9, 13, 30
Exotic Shorthair 75
eyes 68, 71, 73

farm cats 33, 79, 82, *83*
Felidae 6
Felis 6, 8
Felis bengalensis see leopard cat
Felis catus 6, 9, 12; *see also* European wild cat, tabby, blotched
Felis chaus see marsh cat
Felis manul see Pallas's cat
Felis margarita see sand cat
Felis silvestris see wild cat
Felis silvestris libyca see African wild cat
Felis silvestris ornata 12, 13, *14*
Felis silvestris silvestris see Scottish wild cat
Felis torquata 12; *see also* tabby, striped
Felis temminckii see golden cat
feral cats 8, 79, 82, 87, *87*, 88, *88*, 89, 90

Gaston Phoebus, Comte de Foix 43
Gayer-Anderson cat *32*
Gesner, Konrad 46
Gizah 37
golden cat *12*
Gray, Lord 52
Greece 51
Gwentian Code *see* Hywel Dda, Laws of

Harappa 26
'Harleian Bestiary' 47
Havana Brown 75
Herodotus 36, 38
Himalayan 75
'The Hunting Laws of Cambria' *see* 'The Nine Huntings'
Hywel Dda, Laws of 41, 43

Isle of Man 62, 82

James III, King 52
Japanese Bobtail 75, *85*
Javanese 75
Jericho 26
jungle cat *see* marsh cat

kinky tail 64, 84
Kipling, Rudyard 19, *19*
kittens 15, *15*, 16, *16*, 17, *17*, 79, 82, 91
Korat 75

Lachish, Israel 39, *39*
Langland, William 52, *52*
Latimer, Bucks 41
Laud, Archbishop 61
leopard cat 12, 13
'Lindisfarne Gospels' 41, *41*
Linnaeus 6, 9
lion 6, 61
Livre de la Chasse 43, *45*
long-hair 62, 84, 85
Lullingstone, Kent 41
Luttrell Psalter *43*
lynx 6, 26, 43

Maine Coon 75
Malayan 75
Manx 61, 62, *62*, 74, 75, 84
marsh cat 37, 38
Middle Ages 41–6
mouse 33, 36, 40, 52, 70, 82, 90
mummies 36, *36*, 37, *37*, 38, 57, 67

National Cat Club 74
Nebamun 31
'The Nine Huntings' 43
Norwegian Forest Cat 75

Old Cleeve, Somerset 49
Orient 33, 83–5
Oriental Shorthair 75
Oriental Spotted Tabby 75, 77

Palaikastro, Crete 39
Pallas's cat 12, *13*, 61, 68
Panthera 6
'Pasht' *see* Bastet
Pekingese dog 61, 68, *70*
Persian 12, 61, 62, 64, 68, *68*, 70, *70*, 74, 75, 84
Petrie, Flinders 37, 38
Pliny 40, 46

point-colours 64, 66
polydactyly 80, 85
Popopoulos, Greece 40
Puss in Boots *54*

Queen Mary's Psalter *46*

rat 33, 36, 40, 52, 57, 71, 82, 90
Rome/Roman 30, 33, 40, 41
Russian Blue 70, *70*, 74, 75, *75*

sand cat 12, *13*, 70
Sandtun, Kent 41
Satanism *see* witches and witchcraft
Satirical Papyrus *34–5*
Scottish Fold 71, 75
Scottish wild cat 9, 12, 17, 39, 62, 87
short-hair 63
show, cat 61, 74, *74*
ship's cats 90, *91*
Siamese 12, 61, 62, 64, *64*, 65, 66, *66*, *73*, 75, 83, 84
Silchester 30
Singapura 75
skulls 26, 30, 37, *37*, 41, 68, *70*, 71
Somali 78
Spanish Armada 62
Speyside 89, *89*
Sphynx (Hairless) 75, *76*
Steinlen, Théophile *60*, 67, 78, *90*
stubby- or stump-tailed 71, 84

tabby 30, 39, 59, 67, 74, 80; blotched 9, *10*, 12, 59, 61, 74, 80, 81, 82, 85; brown *61*; spotted *61*; striped 9, 12, *24*, 25, *44*, 58, 61, 67, 80, 82, *83*, 85
Tac, Hungary 41
Thetford 41
Topsell, Edward 19, 45, 55, 61
tortoiseshell *see* coat-colours
Turkish Angora 75
Turkish Lake Van 68, *69*

Weir, Harrison 61, 67, 74, *74*
Whittington, Dick 51, *51*
wild cat 6, 8, 9, 12, 13, 39, 43, 45, 64, 66, 68, 71, 79, 82, 87, 88, 90
Winchester Cathedral 49
wire-haired 71
witches and witchcraft 45, 51, 53, 55, *55*